Library of Congress Catalog Card Number: 92-72437

ISBN 1-881753-00-X

First Edition

Manufactured in the United States of America

High Desert Testament

Herdland Eco-poetry: What we had, how we lost it, how we can get it back, and a vision of a world beyond bull.

by Steven H. Rich, Sr.

Author's Preface

What would you do if you knew a wonderful and terrible secret, so wonderful that few will believe its promise, and so terrible that few dare look it in the face? What if it were the kind of secret that hardly anyone wants to know--the kind where to have the reward of its knowledge we must let loose of most of the things to which we now cling for safety? (The very things that the secret tells us we have most to fear, and will certainly destroy us on our present course.)

What would you do if everybody really knew the secret anyway, but very few knew anything about it, and most were using a lot of energy trying to change the subject?

My answer to this strange dilemma is this: clearly, the thing to do is to write a poem; literally, to make a creation; to invite you to come inside and share my thoughts and feelings--to see what I see.

This poem contains key information. The lack of this knowledge has brought dozens, in fact hundreds of civilizations to ruin over thousands of years. The nations of the present day know no more about these things than did the citizens of Ur or Babylonia of old. Our fate will be the same as theirs if we do not act quickly now that this knowledge is at hand. All our achievements in science and technology will be useless and impotent if we do not.

To whomever this work comes, please understand that it comes to you with deepest respect and admiration for your caring and your willingness to spend your time with these lines.

Please read it. Please pass it on. It is intended only as an introduction. The address and telephone numbers where you can get the information you need are on the last page.

You can trust these people. They are not after your money or your mind, but they do need your help. If you have been looking for something meaningful to do to help the environment, this is it.

Like most phenomena associated with Life, this whole thing is abundantly rewarding in every way. Will it make you richer? Yes. Happier? Yes. Healthier? Yes. Sexier? Yup, it sure will! It will even give you a brighter smile and make you a better dresser. (Well, maybe not the last one.)

The message in this poem is true. Our hope lies with Life. Now, that makes sense, doesn't it?

Steven H. Rich, Sr.
Jacob Lake, Arizona 86022

High Desert Testament

by Steven H. Rich, Sr.

When the world was undivided by the many works of man,
Before the fence and highway, and plow and town began,

The plains of the world all swirled with life in pulsing whirling streams, so full and fertile that the truth surpasses all our dreams[1].

Life surged from the plains to the mountains, flowing out through the forests of trees, the air was filled with the flight of wings, life swelled the streams and seas.

This legacy of Life survived until not long ago, and Life's voice still calls

* Definition. Desertification = Loss of biological diversity. Areas of low rainfall can have rich, abundant and diverse plant and animal communities, live water, and live soils. Such are not really deserts but retain the "desert" qualities of openness and stark beauty.

me from the sands and the life of a
high plateau.

There's a place I know on a wild plateau
where a seep of water flows.

There's a little pond just up beyond, and
a sward of sweet sedge grows.

I won't take many people there, just the
ones who can understand, in a quiet
heart, that this place apart is sacred to
their friend.

There's a rock up there, like a frozen
fire, like a sheet of frozen flame. It's
a thing of awe, and there are more,
no two of them the same.

These sculptings stir me to my core,
how the earth, wind, water and sun,
all strove to uncover and shape these
works, so long ago begun.

They spring from out of the living rock,
and of Life they are a part, patinaed
with lichens, dens, and nests, they are
of Life's make and art.

All these rest in a bowl of stone, of
banded subtle hues, and they glow

like they had the fire of life, and it
seems to me that they do.

I can see in their cracks, and joints and
forms, when I have them close at
hand, an expansion of all the shapes I
see in a single grain of their sand.

I have never seen, but I know full sure
that the grains reflect what's in their
smallest parts and each whirling
charge where matter and life begin.

Though most won't grant that atomic
specks have much in the way of
mind, I feel as though we should have
respect for the wit of atomkind.

They know how to interconnect in
ways we depend on, you and me, and
they know how to keep on keeping
on. The little things know how to
be.

When I weary of the ways of men, (and
that doesn't take much to do) I close
my eyes and I'm back again to where
things are staunch and true.

I see in my mind that patch of green,
knowing things of greater worth,

than the monuments grand that
around it stand, lie buried there in
the earth.

The tide of life has ebbed and flowed,
but lately now it seems that a tide of
strife has a hold on life, and it
darkens dreamers' dreams.

I dream of the days when Life's tide was
in, when the ground wasn't bare to
the sun. I long for the sight of the
flocks that flew, and the herds that
were there to run; for the cry of the
wolf and the coyote's voice to ring
through the trees at night, and to
look to the sun as the day is done and
see insect swarms in flight,

To marvel at all life's myriad forms
adapted to this harsh land, and to see
them abound from the eagles on
down, and to see life make soil of this
sand.

In my heart's eye it all transforms, as
the healing Life I see, flowing out
through the draws as the earth
awakes, and I know it can come to
be.

When I was a child the old men told of
grass to the horses flanks,[2] of the way
that it waved as it whipped in the
wind as it stood in its verdant ranks.

I've been led to know how that came to
be, and how it can come again.
Mankind has hold of the tide, the
choice has been given to women and
men.

There's a spirit in man that sees what is,
but our minds can be confused. So
each takes what he can and wants
what's his, and Life becomes abused.

Life was not made to be cut apart, it
exists as a mighty whole. We each
know this in our center heart, in our
cells, in our bones, in our soul.

Men saw Life's Tide, and with greed
they burned and willed that Her
wealth they'd own, but she flowed
from their hands as the tide they
turned and left them to die alone.

So the sun beats down on the barren
ground and I swear that it's by our
choice that its wasted power kills the
hope of soil, for we will not hear the

voice of all that is and that was and will be, of the truth before our face, our eyes are blind for they will not see our freedom and our place.

Even those who wish to be friends to Life, to rescue the land and sea, are so blinded by anger they think that strife will bring what they want to be.

Humans have fought since time untold, almost since mankind's drawn breath.[3] The connection should be pretty clear by now: it's the fighting that causes the death.[4]

Deserts have followed us in our rise and barrenness in our wake. By now it should not be a big surprise that it's humans that deserts make.

But the very idea of a "desert" is garbled in human minds, and much destruction has happened since deserts were misdefined. The words that became our "desert" which is now an unthinking name, could be translated "disconnected"[5] the meaning is the same. We think of a desert as dryness, as parched and sandy plains, as dunes and barren mountains, where it very seldom rains.

But places with no more rainfall than some deserts that we know, are full and alive and fertile, where many creatures grow.

There are deserts as real and as barren as any found on land, composed almost wholly of water, where the dunes of the ocean stand,[6] piled up by the winds like the sand dunes, they move and they roll ever more, with scarcely a life beneath them until they draw near to the shore, where the outwash of the rivers bring nutrients to the foam, or upwelling currents from the deep in the same way give life a home.

The living places, land or sea all have in them combined, water, sunlight energy* and minerals. If we find all these in the right proportions, life's miracle will abound, but if there is

*With the exception of some phenomena on the sea floor involving thermal vents.

too much or too little of these things, less life is found.

Life's creatures are a mixture of the land, air, sun and sea, and are now so interconnected that they cannot separately be.

A "Place of Separation" is what deserts are, and then, where the combination starts is where the desert ends.

Places that once were the wealth of kings, well watered and full and green, are now almost empty of living things, there is little left to be seen.[7]

Men came to these lands and planted fields and hunted for their prey. They took from their wealth but failed to see how they came to be that way.

First they fought with the meat eating beasts and slew the young in their dens. When these were gone, everything had changed and would not be the same again.[8]

The herds that had run in the race for life with the fanged ones every day, were then bereft, and they all behaved in a very different way. They loafed and lazed by the waterside where once they drank and ran, and they stripped the green from that greenest place. Destruction there began.

God's justice plays a mighty joke on the will of the human race: if we rip a creature from the whole, we are forced to fill its place.*

If the running wolves and the stalking cats by man's actions all are gone, we must take their place as shepherds if the whole is to go on.

The swirling dance of the hunters and the hoofed ones brought to be green pastures and still waters in ways we have yet to see.

Their herders gone, the flocks and herds all scattered out alone. The power of their healing wave dispersed, soon they, too, would be gone.

*and do its work.

The key to the Combining in the Herdlands is the herd and the guidance of the herders,without these, life's deterred from bringing the parts together, and the crusted land will shed air, seeds, soil and water, until more or less, it is dead.[9]

Blessed with tools upon their feet to dig and spade and hoe, the hoofed ones make a place for plants to start in as they go.

The power of the herd can smooth and heal the wounds upon the earth.[10] They tend the surface of the ground, their thundering rush gives birth to a million plants in every mile.[11] They tend their greening brood: they prune, they fertilize, they mulch, even as they take their food they bathe each stem with a healing balm, they feed them with their breath, within the Whole they bring the dry lands life instead of death.

Their passing leaves the surface tilled and ready to receive the rain that falls[12] and the eager seeds, and lets the good earth breathe. The creatures living in the soil need air to live and grow. The pounding hoofs like the pounding surf, bring air to those below.

Without the hoofs the soil soon seals, with algae, moss and clay. The air is slow to penetrate, the water rolls away.

A desert starts beneath the ground down where no man can see. The soil life starves for all the things that made the grassland be. Small creatures, those who process food to feed the living plants, are soon without the things they need; dung beetles, termites, ants, and microbes in their many forms, more creatures than we know, make up the teeming soil from which the surface creatures grow.

The Herds brought life to living lands, the gardeners of God, their presence brought the wealth that sprang up from the grassland sod. They made the living springs to flow,[13] they planted living seeds, the outflow of their living filled the other creatures' needs.

I marvel at the cycle: how when one life comes to die, the others come to eat the one that he was eaten by. The grazers eat the growing plants, then in their time are slain by others who go on to die and rise up once again. Their substance sinks into the soil to nourish others who spring up again to vibrant life, shaped as a plant that grew by eating all the eaters and by drawing back their breath, a secret only Life knows: how to make life out of death.[14]

I marvel how the other creatures acted on the herd to give the plant life time to heal. Their different actions stirred the hoofed ones on to move and move and spread their impact wide. Each in seeking food defended all the countryside.[15]

Wherever herds would concentrate and linger for some cause, a guardian would linger too, to guard it with his claws. These guardians existed then in every shape and size, from running wolves to cats, to eagles, down to biting flies. From the waterside to the meadowland to the trees that offered shade, the herds faced peril in the Whole, in every draw and glade. Their favorite feeding places raised up flies out of their dung to drive them off in torment if they tried to stay too long.

The riddle of the path of life the herds paid dear to know. The deaths of millions through the ages taught them how to go. Driven by the weather and the guardians of the ground, they made their way across the land where their needs could be found. Just as the waves of the ocean tide go sweeping back and forth, they followed waves of climate from the south into the north, or up and down the mountains to the valleys as the snow and summer heat determined to their leaders where to go. Each wave of beasts prepared the way for the next wave that would need some action of the others if they were to live and feed.[16]

Each creature is as much a part of the Whole that makes the land as our heart or lungs or stomach or our eyes or arms or hands are parts of our whole body. Why, the very stones of Earth are filled with the parts and breath of Life*, all things have place, and worth.

One is not without the other. Alone we are absurd, without the Whole, the herding beasts will seldom form a herd. They scatter seeking each their own, except when due to fear or mating or migration time they want the others near. Man rarely solves the riddles. We do not have the art. How can we see what makes a herd but never is a part? If we do not perceive the Whole, how can we ever see how one part makes another part and brings them both to be?[17]

So mankind took what they wanted, and slew what they didn't use. The herds went off untended because mankind refused to fill the full role of the herdsman, they were waiting only to kill, but not to chide and to herd and guide, no, those roles went unfilled.[18]

* This is not only a poetic or mystical image, it is literally true. The chemical makeup of much of the earth's crust has been changed utterly by the presence of Life. See Lovelock, James. The Ages of Gaia. New York: Bantam Books.

The plants of the grassland withered, choked with their own dead growth, the soil sealed up, and the soil life starved, and then men said with an oath, "Gods, why doesn't it rain more? and why do we have these floods? It seems like we're either choked with dust or the land is running with mud."[19]

They built dams to catch the water as the springs began to die, but they filled with silt as the land bled soil, still they seldom questioned why. They simply built a higher dam and stronger than before, then mourned in their frustrations as it filled with silt once more.[20]

They filled the range with domestic stock but blindness ruled again, the herdsmen did not guard the grass, they guarded the stock from men.[21] They fought for the best grazing then tried to take their ease, as they swatted flies and wondered why their herds had so much disease, and why the grass doesn't grow so well, and why does it seem so dry? and why do we now have to dig these wells? and why do the calves all die? and why do the cows not calve at all, and why do we now herd sheep? and where did this brush all come from, Dad? and why is this well so deep?[22]

By the time they were down to herding goats and the "drought" went on and on, a real drought came and their crops all died and their flocks as well were gone. They had spent but little time in wondering how things worked before, but they had learned to fight quite well, and so they set out for war.

They marched out over the withered land that they had helped to kill, and left a trail of dead in the sand and more as they climbed each hill. History tells us of their wars, their cultures and their laws, it also tells us that they're gone, but has seldom guessed the cause:

As the ocean tide has cycles, these, too, on the land hold sway, and the dry, low years come in their time and sweep all the fools away. Sweeping them out of the herdlands, or they litter the land with their bones. Men can learn to be part of the whole or they She will never own.

The tale was told over and over, they would ride on down to the coast, and pillage all the people there and make their battle-boast, and earn what they thought was glory. In their need they all were blind, ignoring the pain around them, and the death they left behind, to rule as the Lords of the Rivers by forcing the ways of men, who plotted revenge in their torment till the wheel went round again.[23]

The wonders of the herdlands are not regained with ease. Parts of the world if left to rest will quickly raise up trees, to cover the earth with their leaf-fall, to bind up the soil again, and make a shelter for weak young life from excess of sun and rain. The soil-life returns to the ground there, the water soon sinks in the soil; and all of this happens by nature, and not through men's thought or toil. The growth of the green time falls to earth, the stems rot in little time, and the soil digests this bounty in these gentle, favored climes. They return of their own to abundance, all of Life

does the gardener's deeds, and in
various ways intermingle to fill one
another's needs.[24]

This is not so in the herdlands, with
their long times of heat, cold, and
dry. Grassland may not re-establish,
though hundreds of years go by.
Dead grass stands for years un-fallen,
while the soil-life remains in need,
the spaces between the plants widen,
and yet very little seed of the millions
produced at seed time can find what
they need to grow, and it wastes and
it dies or is eaten, and progress is very
slow, or else can recede still farther if
the herds do not come as before to be
once again God's gardeners, to join
Life's separate parts once more.

Lands once famous for their bounty,
which once yielded to the plow, and
furnished the bread of empires have
become famous deserts now. Before
they were farms they were herdlands,
since before the time of men, who
mined out the wealth of the ages
with the points of the plows back
then.[25]

Since those days much has happened,
mankind has in power grown; and
we think we at last are able to live in
the world alone. We have faith in
our strength and invention; we have
faith in the minds of men, we have
higher dams and deeper wells than
there have ever been. We have no
end of new technology to preserve us
from Nature's yoke, but I tell you I
don't want to be there, when
everyone gets the joke.

It took all our will and our science to
propel us to where we are. Our
survivors will wonder how we came
to miss the point so far.

How we came by enormous labor to
make lands artificially green, while
their dying went unnoticed and their
peril never seen.

It is simply the same ancient story
repeating itself in our fields: the soil
growing paler in color with ever-
decreasing yields. But *we* pour on
the chemical magic, deceived by our
own sleight of hand, our efforts both
comic and tragic, as a desert spreads
under the land.[26]

We work ever harder and harder for what we might have for free. We drill and we blast in the mountains, bore holes in the bed of the sea. While Life's creatures stood ready to serve us, to pour out their wealth in a stream, we spurned them and built our own desert, while lost in a chemical dream. But every dream must have an ending. We'll awake to our barren fields, our petroleum paradise passing, as the desert stands revealed.

It is not now too late to start over, to return to Life's warm embrace, to stand in our place intended, to look the truth full in the face. We have done much to complicate things; cities, roads, fences and more, stand athwart the old paths in the herdlands, where the herds had roamed before.[27]

The same has been done in the woodlands, the patterns of Life are changed. We have battled against Life's creatures since from Life we became estranged.

Our comfort and our convenience, our
notions about our health, quick
yields to convert to money, our
notions of what is wealth, our fears
for our life and safety, our fears for
our stock and grain, have blinded us
to the patterns, and sent us to war
again.[28]

Let any supposed interest be threatened
with loss or harm, we'll spray poison
on rangelands and forest, we'll spray
it all over the farm. To fight with an
unloved creature, we will labor to
dusk from dawn, and we'll poison
the blood of the cattle, and poison
the soil of the lawn.

We hardly have cause to wonder about
water not fit to drink, with our
faultless green grass in the dooryard
that gives off a poison stink, or that
insects we sought to master now take
of our harvest more; as they multiply
faster and faster than they ever did
before, because in our poison passion
we killed off the ones, you see, who
controlled them for years uncounted,
and did all that work for free.

With our miles of monocultures, of vegetables, corn or wheat, each one a mighty banquet of what crop-eating creatures eat, we ought not to be too astonished when a few billion come to stay in our perfectly planted bug hotel that we purposely planned that way.

Since most of the corn was planted with intent to be fed to cows, it seems we might save ourselves trouble if we did some thinking now. We could save ourselves fuel by the trainload, sprays by the megaton, and mountains of fertilizer, if we did what was formerly done, but substitute cattle for bison, let the food chain be more direct. We'd produce as much beef as ever, at more profit, I expect. Under that process Life would flourish, and produce real and lasting wealth, and farmers could live in plenty, without marring their children's health.[29]

We must learn about Life's secrets. They stand there, quite plain to be seen. Life is a complex creature, She never was a machine,[30] and She seeks to preserve her life now, as She waits for us to awake. Though She loves us and needs us badly She has had all She's going to take. When we seek our selfish interests without the whole in mind, she just slips away in sadness, and kicks us in the behind.

Saying "Wake up, my foolish children, it's time that you earned your keep. This glorious world has been dying while your souls have been asleep. Your work is undone, unfinished, it is time now to face your test. The wastelands are unreplenished, and much that was full is waste. Of all matters vast existence, in all of creation wide, only you have the power of choosing, it is up to you now to decide.

"Will you choose the way of the living? Will you let other creatures live in a way that in taking their living they yield all they have to give? Will you foster the mighty cycle, that makes one Life of sea, sky and earth? Will you rise above fear of the unity of the forces of death and birth? Are you ready to heal the deserts of the land, sea, soul, and mind? Are you ready to live in plenty? Will you value the peace you find?

"Man was raised by the Whole for a purpose, raised up with seeing eyes, and a heart that can comprehend the truth and distinguish it from the lies. Do your hearts understand what is coming? Can you see now? Do you hear? Do your souls perceive the power of Life or are you yet slaves to fear?

"Will you fight over what's remaining, or join the severed parts, and make of the world what it can be, as you know it can be in your hearts. Will you wield the power of Life to heal the withered and wasted places, and heal the withered and wasted hope that has left people's eyes and faces?

"The healing of one heals the other. Your growing will let Life grow, if you are willing to see what you see and not what you think you know.

"It is time to be fully human. Own up to your errors and deeds. Cease looking for 'drugs' and scapegoats. In that nonsense lie the seeds of destruction as surely certain as any you care to name. If you miss the way that leads to Life all the other roads end the same.

"In true addictive fashion, the techno-addicts now, in a typical misdirection ploy are blaming it all on the cow.[31]

"'Cows,' they now say, 'are the problem, if we just had no cows we'd be set. The grass would grow the streams would flow forever, and better yet, the greenhouse effect would disappear, global warming would end, it's true! Wild ungulates never belch methane, like these bilious bovines do. From cattle tracks in the cryptogams we'd be forever free, and there would be no cattle dung from sea to shining sea!'

"Such stuff hardly merits a response, but as courteous cows, they do respond the behavioral, animal way, by looking back at you. You can read in their stance and their posture, and their age-wise cattle eyes, that they suffer little confusion as to where the problem lies.

"They've been faithful for years unnumbered to their part in Life's work and Life's dream, and to only those lacking that vision do these worthy creatures seem as other than part of Life's great Whole who are willing their work to do, while asking the pertinent question, by looking back at you.

"They are ready to live in the blazing heat, to deal with the drought and rain, and trek through the snow in the winter, then do it all again. They'll give you their calves if you need them, they will till and sow and plant, prune, aerate, mulch and fertilize but they won't do what they can't.

"They cannot replace all the creatures you have taken from the whole, (though they'll make the way for their return as their part of Life's Work and Life's goal.)

"They can't bring back migrational pathways, (though they'd try if they were free) and they look to you with your vaunted minds to tell them where to be!

"All of Life once guided their travels, they stand ready to do their task: to stand in the stead of the Bison, and it's not too much to ask, of those who are doing the damage in this time of the age of men, to do what *you* were made for, and make it right again.

"If you learn to know Life's cycle, you can bring it all to be. You can have abundant wildlife and clean water running free. If you learn to reproduce the whole by guiding your herds in their ways, you can have again the glories of the herdlands in ancient days.

"You can have all this and much, much more, or heed the strange suggestions (that come from this morbid fear of cows and the remnants of their digestion) to the tender thoughts of some people, those erstwhile urban chaps, whose dread of cow dung sometime makes me wonder if perhaps their potty training time was tough, and it ... sort of ... scarred their minds.[32] But a cow never suffers from that. She knows that she's leaving a gift behind.

"Life's little people share her view, and it's gratitude they feel, as they scuttle, and flutter, and wriggle right up, for a home, and a drink, and a meal, and to do their part to feed the soil, the job the cow's begun, that finally feeds the growing plants that rise to greet the sun.

"In a similar way I find that it is fitting and not strange, that to save the rain forest from cattle use you must heal the cows' home on the range. If the herdlands of the world produced the bounty that they could, there would be little profit in raising cows where tropical forest stood.

"You *can* untwist the tangled thoughts that fear produced in men. You *can* recover happiness and all that's lost, you can! Free eyes will always find the way that leads to being free. Accept responsibility, and then see what you see.

"All Life depends on humankind's acceptance of its role. Mankind's thoughts will always manifest upon the whole."

And then She turns and starts to go, then stops and says again the message we must learn to be full women and full men, "The healing of one heals the other, your growing will let life grow, if you are willing to see what you see and not what you think you know."

Now, Life has kicked me pretty hard. I'm kind of stiff and sore. But in some things I find she doesn't kick me any more. My mind's untwisting in some spots, and the more it does, I find that being where Life has things straight sure helps me to unwind.

In my mind I go to that wild plateau, on the rim I sit and think, and my thought flies wide through the countryside out beyond that awesome brink to where the Lord of Life has hidden safe from the acts of men, the parts of the paradise to come, when the tide flows in again.[33]

There is full rich soil at cliffside springs with the smells of living sod, and the beauty of each setting there would lift the heart of God.

Between the boulders wedged in many a
hidden canyon floor, there live the
plants that once swayed and danced
in the winds of the vales before.

One stunted cottonwood holds on
where many others stood, and sends
out seed hoping for the day the
whole wash grows cottonwoods.

Like the tidal pools at the ocean's edge
these places still can hold the full
strong life of the flood tide that the
Whole land held of old.

When Life's tide flowed up the
mountainsides to the lands of trees
and snows, she left behind these
pools of hope for the desert lands
below.

While the desert creatures held the
ground so it would not fully die, and
for many long years they survived
being left there, high and dry.

These sacred shrines to Life's full
promise send each passing year their
seed to bless the stricken lands when
the time of growth appears.

I'm telling you, *I have seen it come*,[34] when the seeds burst forth anew to grace the ground in the ancient soil where their long ancestors grew! I have seen the water start to flow again in beds long dry, now that the sun no longer boils it back into the sky! I have seen lands fill with birds and beasts, I have seen new meadows grow, in former desert dryer than some deserts you know!

And it all cost nothing but the price most men refused to pay throughout the ages of the world down to the present day: they had to pay attention to the plant life in it's speed of growing so to guide their herd to match the mutual need of beasts and plants. And for that price the land began to bloom,[35] preventing desert's deepening hold, their other certain doom. They had to learn to see the whole, each insect, track and tree, recording changes, days to years. By watching they can see the trends that tell the real effect of their actions on the whole. By this means Life communicates their progress to their goal, or warns them to watch out and change their plans in time to change, so life succeeds to higher Life on their growing, living range.[36]

To do those simple things you pay another price: you must end the endless fighting and begin to learn to trust.

The deserts spread proportionate to man's attention spent attending to which jackass wins some stupid argument.[37] If the attention Life deserves is distracted for too long we just don't see in time to change, and things go very wrong.

I swear that I have seen these things! I've seen the desert heal! And I cannot begin to speak the passion that I feel to see that healing cross the globe forever and abide: my whole soul longs to be there at the turning of the tide.

On that wild plateau I walk among some junipers that grew before the white man sailed here and I think of times they knew.

When I see their foliage thinning from the ground sealed up so tight, and I know their waning's from a cause so easily set right. When I watch flood waters rip and gouge and tear away the land, and carry with them tons of life along with tons of sand, I'm filled with sadness as I watch that hope all rush away, and I know the ground may be full dry within another day, and I feel so much frustration and impatience to proceed, to get someone to listen, to see, to learn, to heed.

It does not have to be this way. Will *you* listen? Will *you* look? Will you go and see it for yourself? Will you read one single book? Will you gain some skills, if you need to, so you know what it is you see? And then put on your working clothes and do this job with me?

When I see floods and dust storms I feel lonely, robbed and sad.

When folks won't give the truth a chance I feel betrayed and mad.

But mostly I feel full of hope and grateful to my core that I know what to do because I just can't feel much more.

Maybe you have a place somewhere,
some place you really know and care about the way I do that windy, wild plateau.

A place that you would fight for--but
now there's no need to fight.

There is just the need to *hear* that place
when it calls you in the night.

The call is in Life's language. Let that
call in you abide.

Then rise in strength to do your part, in
the Turning of the Tide.

Notes

1. The plains of North America held over 100 million large ungulates alone, ranging from the plains bison to wapiti or elk, to mule and whitetail deer and pronghorn antelope and bighorn sheep. Added to these were millions of predators, plains grizzly bears, wolves, coyotes, foxes, badgers, etc., as well as billions of smaller mammals such as prairie dogs and rabbits, mice, etc., and their predators. There were huge numbers of

waterfowl and other birds, plus reptiles and amphibians. Insects were prolific. The soils were rich, deep and productive and the rivers contained diverse aquatic life in large numbers.

The life of the African plains survived intact in places well into this century. Most people have seen photographs of African wildlife filling the land from horizon to horizon. Early European settlers were greeted by even more astonishing abundance: in southern Africa, settlers' ox teams were trampled to death in their harnesses by migrating springbok, an antelope the size of a pronghorn. The plains of Europe and Asia, South America and Australia also held huge numbers of animals, abundant plant life, fertile soils and efficient water cycles.

The reader is encouraged to research the topic if s/he has questions or doubts about the author's claims.

2. All the old cowboys agreed on this point. The author was told many times, "When you went visiting you didn't have to clean your boots because the grass would do it while you rode." In *Sagebrush and Sixshooters* (Rider, Rowland W. Provo, Utah: B.Y.U. Press), page 42, Rider casually mentions that the grass in Houserock Valley between the Paria and Kaibab plateaus was "stirrup high" in 1909. On page 47 Rider, who was employed by the Grand Canyon Cattle Company and well acquainted with their operations, mentions that the Bar Z (as the aforementioned was known) grazed "at least 100,000 head of cattle" on an area consisting of the Kaibab Plateau and Houserock Valley. Less than 1/50 that number are supported there today. Even if Rider were high in his numbers by two thirds the range would have been many times as

productive of biomass a few decades ago as it is today. Whether those numbers could have been supported on a sustained basis is not the point. (Certainly there was extensive overgrazing.) The fact that such numbers were supported for several decades is proof that the resource base was once vastly more prolific and successionally advanced than it now is. In the author's travels he has made a point of interviewing stockmen about old timers' descriptions of the range in early days and has found this memory of an almost unbelievably productive landscape in the late 1800's and early 1900's to be universal. Everyone tells the same story.

3. It is the author's belief that the primary, principal, basic cause of desertification is humans' abuse of one another. He believes that the hurt, anger and confusion resulting from this abuse results in humankind's attention being distracted and our energy diverted from our primary task of learning about and caring for the Whole of Life which includes humans. (In his dissertation "Man and the Soil" [Mickey, Karl B. Chicago: International Harvester Corporation, 1945] Mickey, in the section "Erosion as a Social Symptom," makes essentially the same point, citing several authors.) Certainly the percentage of the sum of human effort used in this task is almost inexpressibly small. Human actions and interrelations with life are therefore often inappropriate and destructive. Of these destructive actions none has done more damage than the distortion of the way herbivores, herding ungulates especially, relate to the environment.

4. Fighting and death have many other well known connections. The connection is deeper and

more profound than we suppose. Those familiar with the Tavistock model of group behavior will recognize the role of fighting in task avoidance. See 3 above.

5. Webster's *Third New World Dictionary* states the derivation of desert to be: *de* (to remove from, or separate) plus *serere* (Latin, to join together) To remove from combination, or disconnect, or take from series, or separate from connectedness are all accurate translations of the combined words from which the Latin *desertum* was compounded. This interesting little aside is not intended to prove anything. The important point is that life flourishes where the water cycle, mineral cycle and energy flow are brought into close connection.

6. For the purposes of this poem the author mentioned only one cause (lack of nutrients) for the comparative lifelessness of many mid-ocean areas.

7. See *The Conquest of the Land Through Seven Thousand Years*. Lowdermilk, W. C. U. S. Department of Agriculture, Soil Conservation Service.

From the evidence one can easily conclude that armies only change governments: agriculture destroys nations. (Agriculture practiced without understanding and caring for the whole.)

8. In the natural world the actions and presence of predators protected riparian areas, meadows and the like from overuse by herbivores. Herbivores need water and also prefer these lush sites for feeding areas. This made them productive hunting sites for predators, which prevented the herbivores

from lounging and loafing in such places as protected livestock tend to do. Alan Savory was the first western scientist to identify the key role of herding ungulates and their predators in what he calls specifically "brittle environments." (Those with generally low, unevenly distributed atmospheric moisture, slow decay rates, and often, temperature extremes. See Savory, Allan. *Holistic Resource Management*. Covelo, California: Island Press.) Without predators, ungulate behavior changes drastically. In addition to loafing in and damaging riparian areas and meadows, they seldom run, thereby failing to trample plant material to the soil surface. They place their hooves carefully while feeding and do not break up the soil surface where it is heavily sealed or "capped," to use Savory's term. They also spend most of their time in small groups or alone rather than forming herds for protection of young and mutual surveillance. Thus "herd effect" is seldom produced, dung and urine are widely scattered, overgrazing is greatly increased in some places while other sites dwindle for lack of ungulate use, soil surfaces seal causing a critical loss of water during heavy rains and preventing the effective aeration necessary to higher succession plants, etc., etc. The whole situation becomes severely distorted.

9. If the reader wishes to know what happens to brittle environments without herding ungulates and their predators or herders in sufficient numbers to impact the entire area, one has only to visit Canyonlands National Park, Chaco Canyon National Historic Site, or any "desert" area where livestock have been excluded for twenty years or more. On a personal tour of such sites, the author and Christopher McKellar of Salt Lake City

conducted pace frequency transects in areas heavily used by livestock and in immediately adjacent "protected" areas. The results were alarming. Sexual reproduction by grasses had practically ceased in the Park sites: plant interspaces (between perennial plants) exceeded three feet. Insect, small animal and large animal activity was markedly less in the ungrazed areas, biodiversity much reduced, and the adult grass plants dead or dying. (Only 16 per cent were in good health.) By contrast the grazing land averaged perennial plant spacing of less then three inches, had many seedlings, and 80 per cent to 90 per cent of the grass plants were in good health. The management of the grazing land was not ideal, and a potentially greater contrast is possible with time controlled grazings, applied herd effect and higher animal impact. (See *Holistic Resource Management* cited in footnote 8 for detailed definitions.)

10. Herds of large ungulates are a powerful force in shaping the earth's surface. The author has personally seen many, many examples where sheep, cattle, bison, deer and elk activity has smoothed and terraced whole mountainsides. The banks of the Little Dolores river on the Mountain Island Ranch had straight vertical banks four feet high until the management (under Al Tohill of the Grand Junction, Colorado, office of the B.L.M. supervision) applied herd effect and animal impact thereby producing gently sloping banks on which willows and other vegetation subsequently grew. This new vegetation attracted beavers which successfully colonized the whole stretch of river. Heavy trampling by ungulates and other large animals is nature's only effective method of

smoothing and healing rills, arroyos, washes and riverbanks.

11. Tests conducted by the author using animal impact, herd effect and grazing to improve aeration, increase water infiltration, till the soil surface, and to plant seeds produced dense seedling reproduction of grass and other species (often in excess of 100 plants per square foot) over thousands of acres. Control areas produced little or no reproduction. Many others of the author's acquaintance have produced similar results. The results of these tests exactly followed Savory's predictions. See *Holistic Resource Management* cited in footnote 8. Bruce Roundy did research on methods of seedling establishment--mechanical and natural--and found that heavy trampling by cattle or other hoofed animals was far more effective than any of the mechanical systems for establishment of grass seedlings. Roundy, Bruce. "Arizona Farmer-Stockman," July, 1990. Tucson: School of Renewable Natural Resources, University of Arizona.

12. The presence of herds under a regime of time controlled grazing (timed to limit plants' exposure to grazing by livestock to one grazing per exposure and to allow grazed plants to completely recover between grazings) has been demonstrated by the author and many others on several continents to greatly increase biomass and biodiversity in "brittle environments." (Again, those with generally low, unevenly distributed atmospheric moisture, slow decay rates, and often, temperature extremes.) See *Holistic Resource Management* "Missing Keys" section.

13. Tests by the author showed soil disturbed by cattle to absorb and retain over 600 per cent more water than that absorbed and retained by "capped" soil. The author simulated rainfall events ranging from light showers to thunderstorms with sheet flows, and simulated a hot, dry wind using an electric heater/fan apparatus. Samples of equal volume from each plot were then weighed, dried, and weighed again to determine water content and to assure an equal weight of soil for the control and experimental samples. The relative protection from wind which the lower parts of the roughened surface received seemed important, as did the tracks' ability to hold water. This greatly increased volume of water in the soil flows downward (if not used by plants and transpired) and recharges underground aquifers, renewing or producing springs in places with favorable geology.

14. This partially describes the mineral cycle found in nature. Colin Fletcher, in his insightful book *The Winds of Mara* (New York: Alfred A. Knopf, 1973, p. 218), beautifully describes this cycle down to the level of dung beetles in the Mara Triangle in Eastern Africa.

"As I drove away I passed close to the bones that were probably all that now remained of the zebra. All that remained, that is, in its old macro-structure. The rest had already been redistributed. The only thing left to be done was for the hyenas to return, after the lions had finally gone. The hyenas would crunch the bones and swallow and digest them and in due course recycle these last remnants of the zebra in their characteristically white, calcium-rich droppings--just as the rest of that recently alive and pulsating animal was already in the process of being recycled to the savanna by lions and jackals and vultures and marabou storks and ants and dung beetles. It was an astonishingly efficient business, this cooperative scavenger service. So efficient that, in spite of the savanna's teeming thousands of living animals, you almost never saw a carcass that had lain long enough to rot and stink and spread disease. So efficient that even stray bones were a rarity."

This cycle proceeds into the soil itself where nutrients are processed by soil insects and other invertebrates as well as microorganisms. These finally mineralize the nutrients and make them available for plants. The great majority of nutrient cycling of plant matter in brittle environments is accomplished initially by ungulates who deliver the material to the soil surface in the form of dung and urine. Additional large amounts are trampled to the ground by excited ungulates. Once the plant material reaches the soil surface the smaller organisms take over and complete the process.

Part of the critical role of ungulates in brittle environments is to get the mass of organic material produced during the growing season onto the soil surface through trampling and 'dunging.' In total tonnage, by far the largest portion of the mineral cycle is accomplished by the soil digesting dung and trampled material.

15. See 8 above. See also *Holistic Resource Management*, "Missing Keys" section.

16. Herbivores in Africa and elsewhere tended to migrate in waves. Those which were able to utilize the coarsest forage came first, followed in succession by smaller animals which required a much higher quality diet. The latter groups often

required high protein regrowth stimulated by the grazing of the larger ones.

17. See 8 above.

18. Predators (as a group, nocturnal an diurnal) in nature perform their functions 24 hours a day all year long, their frequently unsuccessful hunts being just as important to their role as their successful ones. When mankind removes organisms which compete with men for food, men are generally willing to perform only the eating part of the lost organism's role, but leave its many other functions undone. These unfilled functions often spell difficulty and/or disaster for the affected environment, as do the changed relationships.

19. Read the newspapers in California (1991 is a good year for this) or practically any western state.

20. This silting up of dams has been the common experience of humankind worldwide where the protective cover of live plants and plant litter has been removed from the soil. If the dam is considered vital it is usually raised in height, requiring a much larger structure. But if soils in the watershed above the dam are unprotected, the larger dam suffers the same fate. See "Arabia's Frankincense Trail," "National Geographic", October 1985, p. 499.

21. With the notable exception of some migratory herding peoples (Turkhana, Masai) who moved their stock over very large areas and whose grazing practices simulated those of herds in relationship to predation, standard human practice has been to divide the land into units which could

be successfully defended against other humans and try to make a living there from livestock. This led to inevitable distortions of plant-animal relationships and caused desertification. Some researchers believe that the forced settlement of these peoples (the Turkhana, etc.) by governments is responsible for many of the problems in the African Sahel.

22. Poor fertility and disease in livestock as well as loss of live water and falling water tables are all symptoms of progressive desertification. As the desertification process proceeds, herding peoples change classes of livestock from the usually preferred cattle to those which can live and reproduce in less favorable circumstances, goats and camels being the usual final stage.

23. The reader will no doubt be aware of the great tragedies in Ethiopia and throughout the Sahel during the 1970's and 80's. Much of the suffering and death were caused and/or exacerbated by the civil wars in Somalia and Ethiopia. Another account published in the 1930's reads in part: "Once we came upon an abandoned Sart village, where newly thrown-up dams and uncompleted excavations bore witness to the departed population's desperate struggle to retain the vanishing water. In the last extremity of famine the people had thrown their most indispensable possessions on their remaining horses and donkeys and left their homes." (Haslund, Henning. *Men and Gods in Mongolia*. New York: Dutton, 1935 pp. 106, 110.) The ancestors of the Etruscans and Greeks were driven out of Asia Minor by drought. These migrations/invasions by plains tribesmen are an historical commonplace; the Moguls in India and the Manchus in China are examples.

One author sums up in this fashion: "The patriarchal order lends itself to equally impressive abuses. It is nomadic, its hero the wandering Odysseus, or knight errant,...not the city but the Golden Horde...that sweeps down upon the soft and sedentary cultures of the coast and the river valley. It's gods are sky gods with the raging sun at their head. Its depredations are not from decay but by fire and sword. As predatory and greedy as the matriarchy, it cumulates wealth not by unquestioned immemorial custom but by sacred self-serving laws. The perennial routine calls for the patriarchal tribes of the mountains and steppes to overrun the wealthy and corrupt cities of the plain only to be absorbed and corrupted by them in turn...." (Nibley, Hugh. *Old Testament and Related Studies*. Salt Lake City: Deseret Book Company.)

The impetus behind these migrations was desertification (and greed.) See Lowdermilk, W. C. *Conquest of the Land Through Seven Thousand Years.* U. S. Department of Agriculture, Soil Conservation Service.

24. Allan Savory calls these "non-brittle" environments. The eastern and northwest coasts of the United States are examples, as are rain forests, and the climate of northern Europe. See *Holistic Resource Management.*

25. See 23 above, particularly *Conquest of the Land Through Seven Thousand Years.*

26. Though mention of this loss of soil, health and biodiversity will produce heated denials in some circles, it is well know known and acknowledged by academia and government agencies. The Soil Conservation Service will

usually be pleased to supply the reader with information. The Department of Agriculture now classifies farm practice into "standard" and "sustainable" agriculture. Present practice is well known to be unsustainable. Its continuance is a tremendous, culture wide exercise in denial.

27. The situation is far from hopeless. Holistic Management can be applied to any size land or water area, including a flowerpot or a fishbowl. (You may have to learn to cooperate with your neighbors, however.)

28. The use of the "cause and effect testing guideline" of the Holistic Resource Management model demonstrates the futility of standard agriculture's war on "problem organisms. Is the present whitefly invasion of Southern California farmland caused by a lack of pesticides in nature? Is the cause not instead that those lands have been so desertified (stripped of biodiversity) that the white flies are unopposed in their depredations? Further damage to the biodiversity of the land caused by toxic sprays only increases the likelihood of further infestations.

29. It would be easier, simpler and less expensive if cattle harvested forage directly from the land. Andre Voisin demonstrated years ago that grass can outproduce row crops in financial return. (See Voisin, Andre. *Grass Productivity*. Covelo, California: Island Press) Human creativity directed through the Holistic Management Planning Model can solve what is now a tragic waste of resources. (See *Holistic Resource Management* for the Model.)

30. Much has been written in recent years about the errors of the Newtonian/Cartesian world view. Our culture nevertheless continues to make the mistake of trying to impose mechanical "solutions" on organic processes.

31. Cattle and livestock in general are being blamed for everything from ground water pollution to deforestation to global warming. All these are the effects of human distortions of the natural order. Cattle are not allowed to decide by themselves where to live except on the most limited basis. If managed properly they are a powerful tool for the restoration of biological diversity to many environments, brittle ones especially. See 8, 9, 10, 11, 12 and 13 above.

We are addicted to our mechanical culture and therefore look for blame elsewhere.

Livestock can greatly aid the struggle against global warming by depositing carbon in the soil of the vast acreages of the world's rangelands, thereby removing it from the atmosphere. Soils rapidly build up organic content under Holistic Management grazing practice.

32. Rural people as a rule do not share the revulsion some urbanites have for cow dung. They see it as fertilizer and know it to be harmless under normal circumstance. The reader may be interested to know that cow dung creates very little odor unless the animals are unnaturally confined and fed very rich diets, as in feedlots and dairies.

33. If one looks carefully one can find remnant populations of the plants and animals (macro and micro-organisms) that once occupied much of the desert lands of the world in and near the deserts

themselves. These remnants have the ability to repopulate the damaged areas if presented with a suitable successional environment. Xerophytic (desert) plants also have the ability to grow larger and be more fertile under improved conditions.

34. The author has personally seen all the phenomena that Allan Savory predicts in relation to brittle environments. When livestock are managed according to Holistic Management practice, biodiversity increases, water tables rise, aquifers are recharged, soil cover and health increases dramatically. (On the author's land, perennial plant cover increased from 1/2 per cent to 14 per cent in two years, and is now exceeding 30 per cent in places. This is in an eight inch annual rainfall area.) As the process of revitalization continues, a shift occurs toward mesophytic plants. The author has seen this phenomenon in several places: Spurlock Ranch at Navajo, Arizona, Deseret Livestock and Metzger Ranches are notable examples. He has seen published reports of the same thing elsewhere. The former Xerophytes (desert plants) increase in size and vigor and spread into former bare ground areas. In some cases meadows form along watercourses as water tables rise. On the Deseret Livestock and Metzger Ranches, (the former in Utah, and the latter in Arizona) roads had to be relocated because the wet ground would no longer support vehicles. Earthen water holes which formerly filled from surface flooding now fill from ground water which rises from the bottom of these tanks.

Wildlife thrive under Holistic Management. Their needs are planned for, and forage quality improves markedly over that created by other management schemes. Most practitioners of

Holistic Management experience large in-migrations of wildlife, which "vote with their feet" (or wings, or whatever) as to what kind of management they prefer.

In the Prineville, Oregon, BLM district, perennial flows have returned to streams which had flowed only in springtime and are now producing trout. New springs and streams appeared on the Deseret Livestock Ranch as a result of time controlled grazings which greatly improved vegetative cover and opened soils to infiltration by water. Large areas of sagebrush drowned as water tables rose, creating new meadowland.

On the Deseret Livestock Ranch, manager Greg Simons reports that scientists have discovered four new, previously unknown plant species. (The author heard Simons report this at an Arizona State Holistic Resource Management annual meeting in Phoenix, Arizona, in 1988.) The reader is encouraged to contact the Center for Holistic Resource Management for further information. Center for Holistic Resource Management, 5820 4th Street N.W., Albuquerque, NM 87107. Telephone (800) 654-3619, (505) 344-3445.

35. This refers to time controlled grazing. See *Holistic Resource Management*.

36. Holistic Management biological monitoring practice monitors for insect, small and large animal and perennial plant populations, population age profiles, the physical condition of animals and plants, the condition of the soil surface, ratio of bare ground to live plant and litter cover as well as distance between perennial plants plus water and mineral cycle and energy flow data.

These criteria constitute an effective basis for early and effective management measures to progress toward stated goals. If faithfully practiced they do not allow the health and vigor of the land to slip away unnoticed as has been so common in human experience, and as is happening in vast areas of the United States as well as worldwide.

37. See 3 above.

For more information contact:

Center for Holistic Resource Management
5820 4th Street N.W.
Albuquerque, NM 87107
Telephone (800) 654-3619 (505) 344-3445.

Whereas the author advocates Holistic Resource Management, he holds no official position with the Center.

Personal Notes